New York City

TWO HUNDRED YEARS IN MAPS

T0351751

New York City

Two Hundred Years in Maps

from the collection of
Mark D. Tomasko

An exhibition held at The Grolier Club
November 29, 2000 - January 12, 2001

THE GROLIER CLUB 2000

Printed in an edition of 1100 copies.
Designed by The Ascensius Press

INTRODUCTION

I T IS A PLEASURE to display some of my New York City historical material at The Grolier Club. New York was a serious collecting interest of some of the Club's founders, especially William Loring Andrews, a great collector whom I can follow in only a minor way. New York City became a collecting interest of mine thirty-one years ago when the outer edges of lower Manhattan were still filled with small four story buildings from New York's seaport era. In the 1970s almost all of that history disappeared and I became increasingly interested in collecting guidebooks, maps, viewbooks, prints, photographs, and other material that illustrate the growth and development of New York, primarily in the nineteenth and twentieth centuries. While maps were never the sole focus of my New York City collecting, they have always been one of my favorite categories because they are so useful in documenting the history of New York.

The items in this exhibit are a small sampling of my New York City maps and were chosen to illustrate both the development of the city and a wide variety of maps. The cases in the hall contain a brief chronological look at New York over two hundred years, with emphasis on Manhattan. (Until 1898 New York City comprised only Manhattan and part of the Bronx. The consolidation of the City in 1898 added Queens, Brooklyn, and Staten Island.) In the parlor there is a case showing the origins and development of Central Park, as well as cases with a considerable range of other maps and some atlases, usual and unusual.

Space limitations prevented the display of several oversize items, but one of those, my copy of the Dripps map of 1851 (three feet by seven feet) may currently be seen at the Metropolitan Museum of Art in the special exhibit "Art and the Empire City: New York 1825-1861."

I am indebted, as are all collectors of New York maps, to Daniel Haskell's *Manhattan Maps, A Co-operative List*, published by The New York Public Library in 1931. It is a very comprehensive listing of maps of the city, and I have provided Haskell numbers for any of my maps that are in his list. I have also followed his style in this catalog, including dimensions in inches, and horizontal size preceding vertical size. Other references useful for the collector or researcher of New York maps include I.N.P. Stokes' landmark set, *The Iconography of Manhattan Island*, (New York: Robert H. Dodd, 1915-1928.; it has been reprinted several times), and Paul Cohen and Robert Augustyn's *Manhattan in Maps 1527-1995* (New York: Rizzoli, 1997), among others.

While more of my current collecting efforts are spent on other areas, I continue to maintain a fascination with New York City. As William Loring Andrews once remarked about his New York City collecting, it was "the particular hobby [which] I started to ride at the outset and from which I never dismounted."

My thanks to people who assisted me with this exhibit, including Nancy Houghton, Arthur Schwarz, Jean Stephenson, Mary Young and Maev Brennan, who helped with its installation, as well as Mark Samuels Lasner and Eric Holzenberg, for their advice and assistance. Lastly, special thanks to my wife, Nancy Norton Tomasko, who shares my fascination with old (and new) paper, and who helped with all aspects of this exhibit.

MARK D. TOMASKO

November 2000

THE MAPS

A Plan of the City of New-York & its Environs to Greenwich, on the North or Hudson River, and to Crown Point, on the East or Sound River . . . Surveyed in the Winter, 1775. Sold by A. Dury . . . [London]. [Surveyed by] . . . John Montresor, Engineer, P. Andrews Sculp.

Haskell 307. Size: 20⅛ x 25¼. Scale: 5 in. = ½ mile.

Known as the "Montresor" map, this plan was first published in 1766, the date it actually depicts. It was made from a survey done by John Montresor between December 16, 1765, and February 8, 1766. While this 1775 version claims to represent the City in 1775, there are no changes between this edition and the earlier one.

A Plan of the City and Environs of New York in North America. [published in the *Universal Magazine*, London, November, 1776].

Haskell 418. Size: 14⅛ x 11⅛. Scale: 2¼ in. = 2400 ft.

One of several maps of New York published in England during the Revolution. The legend in one of the north-south roads in the center of the island should dispel any doubts as to the national origin of this map: "Road to Kings Bridge where the rebels mean to make a stand."

Map of New York I. with the adjacent Rocks and other remarkable Parts of Hell-Gate. by Thos. Kitchin Sr. . . . Printed for R. Baldwin . . . For the *London Magazine*, 1778. [published in April, 1778 in the *London Magazine*].

Haskell 518. Size: 7¼ x 9⅝. Scale: 2 in. = 3 miles.

This map shows the small settled part of the island (essentially from what is now City Hall south), the major roads, a few estates, and particularly the difficult navigational area at Hell Gate where the East River and Harlem River meet Long Island Sound. Long Island is alternatively called "Nassau" on this map, and Brooklyn is "Brookland."

Plan of the City of New York. Tiebout Sculpt. [published in William Duncan's *The New-York Directory*, 1792-1793, printed by T. and J. Swords, New York].

Haskell 602. Size: 14⅛ x 8¾. Scale: 4 in. = 4000 ft.

The second edition of the 1789 Directory Plan (in 1789 the first map appeared in a New York City directory) which shows the city in the years following the Revolution. There are 38 references on the map, of which half, 19, are for houses of worship. Cornelius Tiebout was an important early American engraver who also engraved bank notes.

Plan of the city of New-York, with the recent and intended improvements. Drawn from actual survey by William Bridges, City Surveyor; A.D. 1807. Engraved by Peter Maverick. Published by Isaac Riley New York 1807.

Haskell 642. Size: 13 x 12⅜. Scale: not given.

This map was prepared to accompany New York City's first guidebook, the *Picture of New York* by Dr. Samuel Mitchill. Many of the streets shown, especially in the northern part of the map, were never laid out as such, as this map was William Bridges' appropriation of the "Mangin-Goerck" plan of 1803. In addition, the shoreline on the northeastern side of the map is mostly imaginary as well, as the island was not built out to the extent shown. Peter Maverick was a leading engraver on copper who also engraved bank notes.

[Manuscript notebook of John Randel Jr., 1812, containing measurements for Manhattan streets], not paginated, oblong 12mo.

John Randel Jr. was the surveyor responsible for the Commissioners' Map that laid out the grid on Manhattan. This notebook contains his work in measuring certain streets and avenues in 1812, including 34th Street, parts of Avenues 1-3, 5, and A-D, among others. A hint of the challenge of doing this work is shown in the entry here, July 11, 1812: "The pegs at 1 Avenue on this cross line are ploughed up by Frederick Rigger[sp?] - therefore take rods to Peg A in Mr. Morris' garden at the end of 14 Rods . . ."

A Map of the City of New York. Published by A.T. Goodrich. Engraved by H. Anderson. Entered . . . 1827 . . .

Haskell 710 var. Size: 28⅝ x 38½. Scale: 5 in. = 2500 ft.

This map was the primary New York City plan used in the late 1820s and early 1830s. Goodrich was the publisher of the main guide to New York City in this era, and a smaller version of this map appears in the guide. According to I.N.P. Stokes in his *Iconography of Manhattan Island*, the "Parade" is only shown on the earliest issues of this map.

Map of the city of New York with the latest improvements. By H. Phelps. W. Hooker engr. & copper-plate printer. Entered . . . 1830 by Humphrey Phelps. [New York].

Haskell 736 var. Size: 19¾ x 16¾. Scale: 3 in. = ½ mile.

The shaded areas denote the settled parts of the city. Note the key, listing three museums in New York at this point: the American, Peals [sic], and the New York. This map subsequently went through a number of editions in the 1830s, and some of the later editions, which presumably used the same plate with alterations, are poor impressions.

Brooklyn Fire Insurance Company [policy of insurance], printed by A[lden] Spooner, Brooklyn, New York, [policy dated 1835].

Map size: 5⅜ x 3¼, on a sheet folded to 10½ x 16⅜.

Shown here is an insurance policy from one of Brooklyn's two pioneer financial institutions, the Brooklyn Fire Insurance Company (the other is the Long Island Bank). The policy heading is decorated with a detailed map of what was then the Village of Brooklyn. Note, however, that the insurance was on the contents of a house in "the City of New York," and besides covering goods such as "household furniture" and "wearing apparel," it included "printed books."

Staten Island, Map of New Brighton property belonging to the New Brighton Association in the Town of Castleton, . . . Surveyed and Drawn by James Lyons 1835, P.A. Mesier's Lith., N.Y., [accompanying *Description of New Brighton, on Staten Island, opposite the City of New York,* (New York: New Brighton Association, 1836)].

Not listed in Haskell (not a Manhattan map). Size of map: 19 x 17½. Scale: 1 in. = 350 ft.

An elaborate prospectus containing a map, text, and view of an early real estate development on Staten Island, before the island became part of New York City. This development adjoined Sailors Snug Harbor, which was being built about the same time. Both the map and view are lithographs; this type of high quality prospectus was probably made more financially feasible by lithography.

Map of the city and county of New York. By David H. Burr, published by Simeon DeWitt . . . Third Edition, 1840, Stone & Clark, republishers, Ithaca, N.Y. [published in

David H. Burr, *An atlas of the state of New York*. Re published at Ithaca, N.Y. by Stone & Clark with corrections and improvements, 1841. plate 2-3].

Haskell 727. Size: 48¼ x 20. Scale: 3⅛ in. = 1 mile.

The city was built up to the vicinity of Union Square by the 1840s. Virtually none of the streets in the Twelfth Ward (which comprised the entire island north of 40th Street) existed. Note the path of Broadway, which becomes the Bloomingdale Road. The grid plan projects not only a Twelfth Avenue on the west side, but a Thirteeth Avenue also. The island was never expanded with landfill that far west.

New York. London: Edward Stanford, n.d. [published in Society for the Diffusion of Useful Knowledge, *General Atlas* (1847?)].

Haskell 842(?). Size: 14¾ x 11⅞. Scale: 1⁷⁄₁₆ in. = ½ mile.

The lower corner vignettes show "Broadway from the Park" on the left (the scene includes City Hall park, Astor House, and St. Paul's Chapel), and "City Hall" on the right. Note that the East River is also called "Strait of the Sound," an uncommon name for that waterway.

Map of the city of New York, Drawn by D.H. Burr for "New York as it is in 1846." . . . Published by C.S. Francis & Co. . . . [N.Y.].

Haskell 775. Size: 10¾ x 12⅜. Scale: not given.

The map itself dates to 1833, and was republished repeatedly by C.S. Francis & Co. for its "New York as it is" guidebooks. This version, published and sold separately with a attractive border of fifteen building vignettes, was likely prepared for decorative purposes. In the lower right corner of the map is a key to twelve public buildings in Brooklyn.

Map of the country thirty three miles around the city of New York. Revised Edition, 1858. Drawn by G.W. Colton. Published by J.H. Colton, New York. Engraved by J.M. Atwood. Entered . . . 1846. . . .

Haskell 1068. Size: 20¾ diam. Scale: 1 in. = 3 ¼ miles.

This map was a derivative of the "Eddy" map of 1812 which showed thirty miles around New York. It was subsequently re-engraved several times, including by J.M. Atwood for the Coltons in 1846, and reissued various times, as late as 1873. The elaborate border and attractive coloring no doubt helped keep it in print. Note the city halls of New York and Brooklyn in the lower corners.

Topographical map of the cities of New York, Brooklyn, Williamsburgh & Jersey City, and villages of Hoboken Greenpoint Astoria. Published by M. Dripps . . . New York, 1854. Entered . . . 1853 . . .

Haskell 1005. Size: 24½ x 35. Scale: 1 in. = 1800 ft.

For Manhattan, 53rd Street was as far north as the City reached in 1853 except for the villages of Yorkville, Harlem, and Manhattanville. The shaded areas represent the populated parts of the cities. The Battery was being extended to surround Castle Clinton at the foot of Manhattan, and the choice for a park was shifting from Jones Woods (East River, 66th to 75th Streets) to the central location, which is also shown in green (5th to 8th Avenues., 59th to 106th Streets).

[Map of New York City from 50th St. to 127th St. showing the proposed Jones Park and Central Park, to accompany Document No. 83, Board of Aldermen (of the City of New York), January 2, 1852.] Lith. of McSpedon & Baker, New York.

Haskell 975. Size: 20 x 15. Scale: not given.

Jones Woods, the tract of virgin forest between the East River and Third Avenue, 66th to 75th Streets, was the original choice for a major park for New York. The Board of Aldermen subsequently reconsidered that choice, and this map shows both the new area proposed, Fifth to Eighth Avenues, 60th to 106th Streets, and the previous choice. Key factors in the decision to choose the central location were that it was more than four times as large as Jones Woods, and had more varied topography (and, in particular, topography that was not very suitable to normal urban development).

Central Park. Memorial of the Common Council . . . to the Legislature. . . . Resolution of the Common Council, directing application to be made to the Supreme Court. . . . Order of the Supreme Court, appointing commissioners, November 11, 1853 . . . [map of New York City between 23rd and 125th Streets, and 112 maps and transverse vertical plans of land included in Central Park. sm. folio, 1855? Copy owned by James F. Ruggles, Clerk of the Commission].

Haskell 1008a.

The condemnation of the land for Central Park took approximately two years, 1853-1855. James F. Ruggles was the clerk of the Commission in charge of acquiring the land, and this is his copy of the atlas of the land taken, with notations next to each building indicating how much was paid in condemnation. On the title page is a brief description of his work, which includes the following: "James F. Ruggles / occupied in business of Central Park Commission from Nov 29 1853 to Dec 15 1855 / appointed Clerk of Commission Nov 1853 / Commission filed Abstract Oct 4 1855 / 619 Meetings held . . ."

Shown here is the largest part (83rd to 85th Sts.) of "Seneca Village," the African-American community which was the only significant settlement in the park lands prior to the building of the park. Its location was from approximately 81st Street to

87th Streets between what would have been Seventh and Eighth Avenues. Most of the houses were acquired for the low to mid-hundreds of dollars, while $2010 was apparently paid for the Free Episcopal African Church, the largest building in the community.

Map of the lands included in the Central Park, from a topographical survey, June 17, 1856. By Egbert L. Viele. [Published in *New York City - Commissioners of Central Park. First annual report on the improvement of the Central Park. January 1, 1857.* New York, 1857].

Haskell 1029, top half. Size: 44¼ x 14. Scale: not given.

Viele was the original designer and engineer for the Central Park, before Olmsted and Vaux and their design supplanted Viele's design and Viele himself. This map shows what was on the land that became the park. Note especially the settlement at Eighth Aveue between 82nd and 86th Streets; this was "Seneca Village," the African-American community that the park displaced. This map is the top half of a larger sheet which carries a second map showing Viele's plan for Central Park.

Map of Central Park New York. Exhibiting the Drives, Promenades, Walks, Buildings, Ponds, Rocks, &c. as far as finished up to July 1859. Drawn by J.P.M'Lean, engraved and electrotyped by A.H. Jocelyn. Entered . . . in 1859 by P. Burger & Co.

Not listed in Haskell. Size: 17 x 3⅞. Scale: not given.

Eight pages of text, constituting a guide, accompanied this map, with another thirteen pages of advertising. Construction of the park had begun only two years before this publication. A portion of the Park had just opened in the Fall of 1858, making this a very early guide to the Park. Wood engraver Albert Jocelyn's map has a folk art quality to it.

Map of the Central Park showing the progress of the Work up to January 1st 1861. Fred. Law Olmsted, Architect in Chief; Calvert Vaux, Consulting Architect. Sarony, Major & Knapp Liths., N.Y. [Published in *Fourth Annual Report of the Board of Commissioners of the Central Park. January 1861* New York: William Cullen Bryant, 1861].

Haskell 1108, No. 2. Size: 34 x 7¾. Scale: 1 in. = 400 ft.

Annual reports by the Board of Commissioners were issued every year while the Park was under construction and detailed the progress to date. By the beginning of 1861, a considerable amount of the Park south of 79th Street had been completed.

Map of the Central Park 1871-2. Olmsted and Vaux, Architects. [Published in *Second Annual Report of the Board of Commissioners of the Department of Public Parks for the year ending May 1, 1872.* New York: William C. Bryant & Co., 1872].

Not listed in Haskell. Size: 34½ x 10. Scale: not given.

By the late 1860s the Park was largely completed. Note the large plot marked "site for art museum" with the original Metropolitan Museum building surrounded by an outline of projected growth. Note also the "Site for museum of natural history" on what was known as "Manhattan Square."

Map of five Central Park lots at the Fifth Avenue and 63d Street . . . estate of John Mason, deceased . . . to be sold at public auction . . . 23d March, 1869, by E.H. Ludlow & Co., Auctioneers. . . .
Not listed in Haskell. Folding card, 11¾ x 10½ open.

A remarkable piece of ephemera showing not only a map of the lots to be sold, but also photographs looking into and out of the park from Fifth Ave. and 63rd Street in 1869. Note how spare the vegetation looked ten years after the opening of this part of the Park, though it is winter. The text notes that the house shown ("Mr. Martin's residence") was the first house "erected on the Park" and that it was designed in the "Venetian Gothic" style by J.W. Mould, who designed structures in Central Park. Note also how the nearest buildings to the east are at Third Avenue.

Topographical map of the city of New York showing original water courses and made land. Prepared under the direction of Egbert L. Viele . . . Ferd. Mayer & Co. lithographers . . . New York. Entered . . . 1865 by Egbert L. Viele . . . [published in *The Topography and Hydrology of New York* by Egbert L. Viele, New York: Robert Craighead, Printer, 1865].

Haskell 1135. Size: 62 x 19¼. Scale: 1 in. = 1000 ft.

This remarkable map is one of the very few nineteenth century maps of the city that is heavily used today. Contractors study it to determine the location of underground water courses in Manhattan. Ironically, its original purpose was related to health. Viele's text accompanying the map is a plea for better drainage and wider streets. He states that " . . . the principal cause of fever is a humid miasmatic state of the atmosphere, produced by . . . excess moisture in the ground from which poisonous exhalations arise . . ." The connection between poor drainage and fever had been made, but the science of the time had not figured out the insect connection, among other things.

[Manuscript ledger of Garret E. Winants, ("*G. E. Winants Real Estate Book 1868*") recording his purchases and sales of real estate, along with family and autobiographical information.] 242p., sm folio.

The ledger is decorated in color with eagles and other drawings by Winants. Garret E. Winants (1813-1890) was born on Staten Island, the son of a ship captain who died when the boy was ten years old, throwing the family into poverty. Winants later became very successful in the shipping trade, and invested some of his money in real estate in New York City and vicinity. He was a self-made millionaire with little formal education, and was very religious and patriotic. He had a talent not only for business, but also for drawing.

This book is an example of how a real estate investor could both keep track of his property and enjoy the real estate when sitting at home. For record purposes, he drew a map of each property he bought, and noted on the opposite page the purchase price, and then later, the selling price, and calculated the profit. The whimsical decoration made it an appealing volume to view.

In addition to the actual title pages, shown here are copies of several other pages, including his purchase of corner lots at Eighth Avenue and 63rd Street for $2265 in 1851. The current site for Central Park (across the street from these lots) was chosen in 1853, and was not the original choice, so this purchase was very fortuitous. In 1860, Winants sold the lots for $34,000 with $6000 down, and in 1862 took back the property with the buyer losing the $6000 down payment. Evidently Winants owned the property at his death, as a note in another hand relates that the lots sold for $106,000 in 1899.

Plan of New York City, from the Battery to Spuyten Duyvil Creek . . . Based on the surveys made by Messrs. Randall & Blackwell, and on the special survey by J.F. Harrison. Published by Matthew Dripps . . .New York, 1868. 19 maps, folio.

Haskell 1182 var.

One of the earliest real estate atlases of Manhattan (insurance atlases had very different scale and detail, and date to the

1850s) from the publisher who issued the first map showing all the buildings in New York in 1851. Note how few buildings there are on the east or west sides of Manhattan from the 60s to the 90s at this time. Only Yorkville, on the east side in the 80s, really has significant settlement.

Bartlett's Illustrated map of New York City or Stranger's guide showing the public buildings, places of amusement & its various architectural features. Entered . . .1870. Designed and drawn on stone with pen and ink by G.H. Bartlett & published at his office . . . New York.

Haskell 1216. Size: 34½ x 20. Scale: not given.

An unusual "pictorial" map showing a number of prominent public and charitable/religious buildings, and many business buildings, some well-known, and some obscure. There is a sixteen page "advertiser" accompanying the map, and, not surprisingly, most every business represented therein has its building illustrated on the map. This is an early example of a map prepared partly for advertising purposes. Surprisingly, there are few pictorial maps of Manhattan (not including "birds-eye view" prints) until the late twentieth century.

The New York Elevated Rail Road Company [stock certificate, proof impression, engraved and printed by the National Bank Note Company, New York, (c. 1870s), accompanied by a die proof vignette of a map of the company's lines]. (See illustration facing page.)

Size: map, 5¼ x 1½; certificate, 11¼ x 7½.

Occasionally a rail road would have the map of their route engraved for their securities. This extraordinarily detailed map of Manhattan done by National Bank Note for The New York Elevated Rail Road Company is one of the best examples of this genre. It is the smallest map in the exhibit and a tribute to the skill of the bank note engravers.

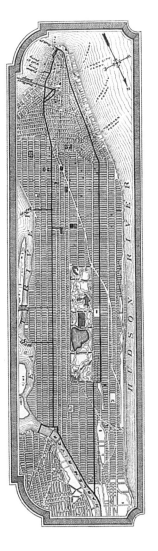

NEW YORK ELEVATED RAILWAY

NATURAL GAS & MFG. CO. N.Y.

Citizens & travelers guide map into and from the city of New York and adjacent places. Published by D.A. Edsall & Co. . . . New York. 1877. . . .

Haskell 1305. Size: 19 x 26⅛. Scale: 4¼ in. = 1 mile.

> This map was designed to provide information on transportation, showing and listing in detail the railroads, the elevated rail road, the horse-drawn street car lines, the ferries, and the steamship lines. By the 1870s the city's growth and significance as a commercial metropolis was such that a good map was necessary both for getting to and from New York and for transportation within the City.

The Hudson River Map via the New York Central & Hudson River R.R., New York: Bryant Literary Union [c.1890s], copyright by Wm. F. Link.

Haskell not listed; probably a 1313 variant. Size: 5½ x 106. Scale: 1 in. = 1½ miles.

> Accompanied by 16 pages of text and views, this map claims to carry the names of 500 "prominent residences and historic landmarks" on the Hudson, from New York City to above Albany and Troy. Fully extended, the map is almost nine feet long.

Rapid transit map of New York City. By J. Bien & Co., New York, 1901. [Sponsored by Marks Arnheim, Tailor]

Haskell 1571. Size: 37 x 12¾. Scale: not given.

> The building of the original line of the New York subway by the Interboro Rapid Transit Co. was a major landmark in the City's history. The project was started in 1901 and opened in 1904. This broadside, published the year construction was started, shows the map of the original lines (even the connection from City Hall down Broadway into Brooklyn, which was

built later), a cross section of the Astor Place station, a cross section of Manhattan, and a panoramic view of lower Manhattan.

Historical Geology Sheet, New York - New Jersey: Staten Island Quadrangle, and Brooklyn Quadrangle, in *Geologic atlas of the United States. New York City folio.* Washington D.C.: U.S. Geological Survey, 1902.

Haskell 1576. Size of maps: 13¼ x 17½. Scale: 1: 62500

Besides providing comprehensive information on the geology of the area, these maps also provide some of the most detailed information on streets and structures outside of the built-up cities, a characteristic of U.S.G.S. maps to this day. At this time Staten Island was almost completely rural, as was much of Queens.

Apartment Houses of the Metropolis, New York: G.C. Hesselgren Pub. Co., [c.1908]. 298p., large 4to.

Perhaps the maps that interest New Yorkers the most are floor plans for fine apartments. This unusual compilation of photographs and plans of major apartment houses was assembled about 1908 by the firm that published many of the prospectuses for such buildings. The Dorilton (71st and Broadway), pictured here, fortunately survives today.

Miniature atlas of the borough of Manhattan in one volume. Brooklyn - Manhattan: E. Belcher Hyde, 1912. 472p., 8vo.

Haskell 1743.

This was the smallest format for the New York City real estate atlases, a size used only in the 1910s. The first nineteenth century atlases showing the lots and buildings were large folios, with the "miniature" and "desk" sizes appearing in the early twentieth century. An enlarged version of the desk size atlas is the only one published today. The pink color indicates brick buildings, and yellow indicates wooden buildings. Shown here is the Grolier neighborhood.

The Dr. John A. Harriss System, American Multiple Highway [photograph of drawing of proposed multiple highway system in New York City, c. 1920s.]

Size: 18 x 14¾

This remarkable proposal ("patent pending") indicates the level of frustration with traffic congestion in New York by the 1920s. Since the elevated railroads were still in existence on several avenues at this time, and the noise and blight they created were quite obvious, it is amazing that Dr. Harriss evidently thought that the city could bury three or more primary avenues and a number of cross streets with this road system.

The Hamilton Aerial Map of Manhattan. Copyright by W.L. Hamilton. Hamilton Aerial Map Service . . . New York, [1927]. Photography and assembly by Hamilton Maxwell, Inc.

Haskell 1970. Size, each sheet: 28¼ x 28. Scale: 1 in. = 200 ft.

The Hamilton Aerial Map was an unusual product. Aviation, combined with photography, gave a new dimension to map making, and produced the ultimate map, an aerial atlas. There were only two such atlases published for New York in this era (or any era, as far as I know)—the one seen here, and the Fairchild Aerial Camera version done in 1924. This Hamilton effort was done on a much larger scale (1 inch = 200 feet) but covered only Manhattan; the Fairchild version covered all five boroughs but had a scale of 1 inch = 1 mile. Today, of course, U-2 maps and a variety of satellite imagery make this kind of mapping more common, but it seldom appears in book form.

Average monthly rent by blocks, Manhattan. Supplement to Survey of the New York City Market. Prepared by Consolidated Edison Company of New York, Inc. and its System Companies. Source: U.S. Bureau of the Census, 1940.

Size: 14 x 33½. Scale: not given.

> Since the cost of housing is closely related to the economic status of the occupants, this map serves as a general indicator of the wealth and poverty of different areas of Manhattan. While this map was prepared primarily as a guide to the New York City market, the information is also useful for civic planning purposes. Maps of this type for New York are uncommon in published form. There was a book version done in the early 1940s for the whole city, published by a consortium of New York City newspapers.

New York. Map of Midtown Manhattan. Compiled, Prepared, and Published by Anderson Isometric Maps, New York, 1980–1981.

Size: 23¾ x 36¾.

> A beautiful isometric rendering of midtown Manhattan done by Curt Anderson in the late 1970s. Anderson and several associates made scale drawings, block by block, to do this spectacular map. The only similar effort was by a German cartographer, Hermann Bollman, in the early 1960s, and Bollmann's map was pictorial, not isometric (that is, exactly to scale).

Mapeasy's Guidemap to Midtown Manhattan, published by Mapeasy, Inc., Amagansett, New York, 1992.

Size: 34¼ x 23, printed on both sides.

> This type of informally drawn plan with all "hand done" lettering is an interesting variation on the traditional tourist maps of New York.

[Heritage Trails New York Walking guide to lower Manhattan], published by Heritage Trails New York, 1995.

Size of map portion on endleaves: 7⅞ x 9.

This guide to lower Manhattan contains in its folding endleaves a very good pictorial map of lower Manhattan which allows the visitor to navigate the area more easily by recognizing the buildings.

New York Popout Map, published by The Map Group, U.K., 1999. Maps copyright Compass Maps Ltd.

Size: two maps, each 9⅞ x 8½.

The 1990s brought some unusual folding maps for visitors to New York; there was an earlier U.S. version of this type of folding map (*New York City Unfolds*), since then used by this publisher.

Manhattan Block By Block, A Street Atlas by John Tauranac, [Santa Barbara]: Tauranac Maps, 2000. 172pp.

Page size: 3¾ x 9.

The most recent pocket set of maps to Manhattan carries a wealth of information on the location of a wide variety of institutions and organizations, in addition to more prominent apartment houses and other notable buildings. Particularly useful are street numbers for every block.

ISBN - 13: 9781605831152
ISBN - 10: 1-605-83115-8

Tomasko/ New York City: Two Hundred Years in Maps